THE GOLDEN LIFE of SALLY SUNFLOWER

By Imani Ariana Grant

Illustrated by Simonne-Anais Clarke & Zachary-Michael Clarke

Library of Congress Cataloging-in-Publication Data

Grant, Imani Ariana

The Golden Life of Sally Sunflower/ Imani Ariana Grant

p. cm.

Illustrated by Simonne-Anais Clarke and Zachary-Michael Clarke

Edited by Tamira Butler-Likely

Summary: Imani and friends attend a plant nursery. Imani meets Sally, who takes her on an awesome adventure where she learns about plant growth and reproduction.

ISBN-13: 978-1-948071-31-4 (paperback)

ISBN-13: 978-1-948071-32-1 (hardcover)

Title I.

2019935730

www.laurensimonepubs.com

Dedication

To my mom, Shaneika Burchell-Kerr,
who encourages me to write
and pursue my dreams.

To the Love of Literature (LOL)
Book Club members.

Imani and her classmates went on a field trip. They visited a plant nursery to learn how plants grow.

Everyone went their separate ways to explore
Suddenly, a small voice said, “Hey down here.”

“Who is that?” Imani replied.

"My name is Sally.
I am the little seed beside your feet. Can you please help me?"
"Sure! How can I help?"

“Put me in soil and make sure I have enough water and sunlight, but not too much!” instructed Sally.

“Ok! I’m on it!”

Imani found Sally a perfect home in her backyard. Every day she checked on Sally.

“Wow, Sally! You grew into a *sprout*. Can you tell me what happened?” asked Imani.

“When you gave me water, you gave me life. My seed became activated and developed *roots,* which anchored me to the soil. My seed also provided the *nutrients* I needed to grow into a sprout. This process is called *germination*. Continue to give me water and my roots will deliver all the nutrients to my *stem* and protect me from *erosion,*” answered Sally.

“Wow, that’s a lot of information to digest!” laughed Imani.

“Hmm, let’s see. So, if I found you at a plant nursery then…maybe you’re a type of flower!” guessed Imani. “What type of plant are you going to be?”

Sally smiled and said, “Let’s figure it out together.”

SPROUT
STEM
ROOTS

Imani kept on watering Sally and waiting for her to grow. A week went by and Sally grew and grew and grew.

She had a long stem, wide leaves, and a beautiful yellow flower. “Wow! Sally you look amazing! I never thought you would be a sunflower.”

"Yes, that's right. I am Sally Sunflower. Thank you for taking care of me!" Sally said gratefully.

"Oh Sally, I have a great idea! Let's take a 'selfie' together," Imani requested.

“Sure! That would be lovely.”

Later that evening, Imani noticed a change in Sally's position and asked, "Sally, why did you move?"

“Sunflowers face the east in the morning and the west in the evening to follow the sun,” Sally explained.

“Why are you following the sun?”

Sun

Energy

“I convert the *energy* from the *sun* into food. This process is called *photosynthesis*. During photosynthesis, the green pigment in my leaves, called *chlorophyll*, helps capture energy and convert it into food. My leaves also take in *carbon dioxide* and release *oxygen* for animals and humans to breathe.”

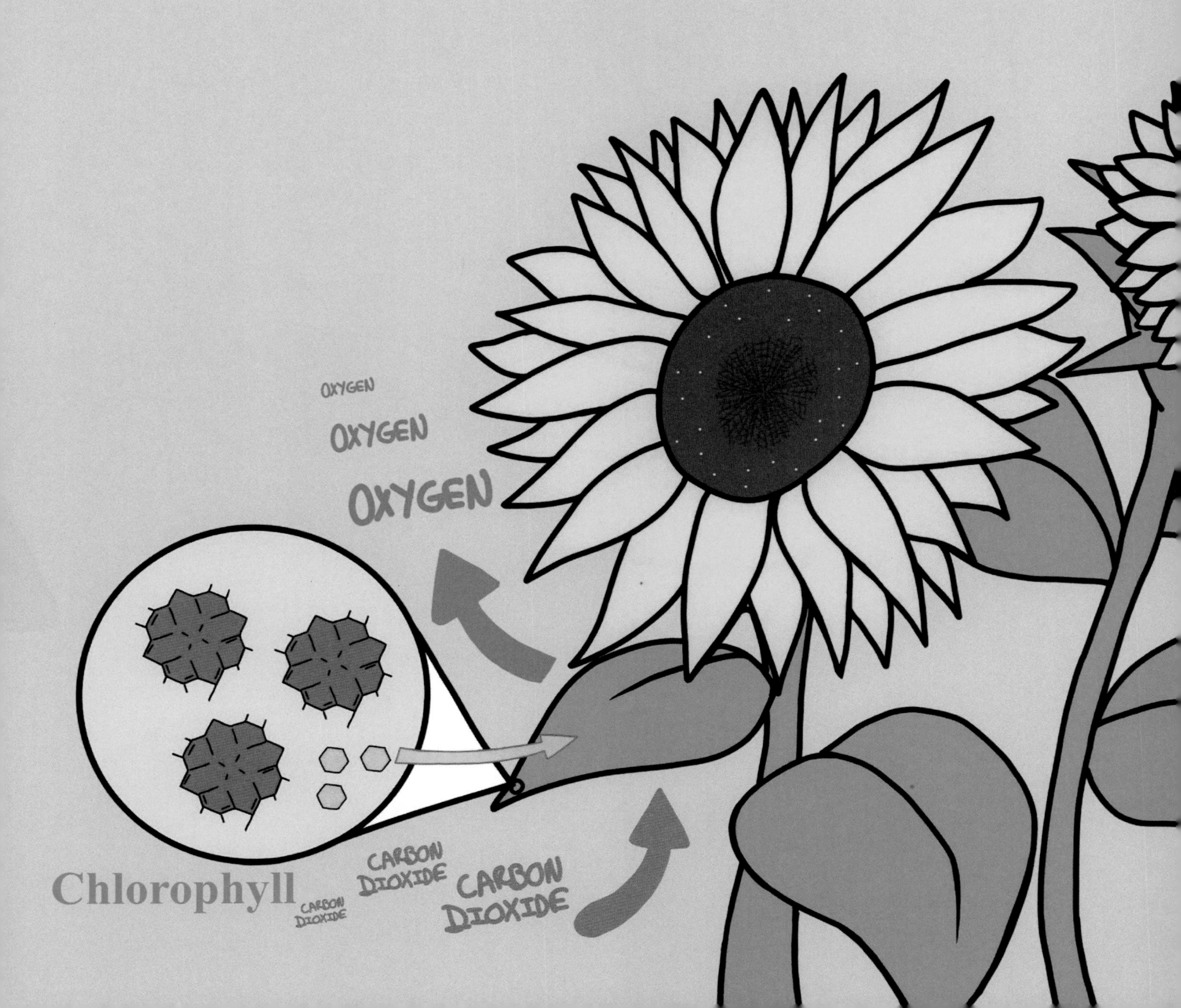
OXYGEN
OXYGEN
OXYGEN
Chlorophyll
CARBON DIOXIDE
CARBON DIOXIDE
CARBON DIOXIDE

Pollinator
Flower
Pollen
Pollinator

“How do I get more sunflowers to grow?”

“My *reproductive system* produces more sunflowers with the help of *pollinators*. When pollinators such as honey bees, bumble bees, butterflies, and other insects land on my *flower*, they transfer *pollen* on their bodies and transport it to allow *pollination* to take place,” Sally explained.

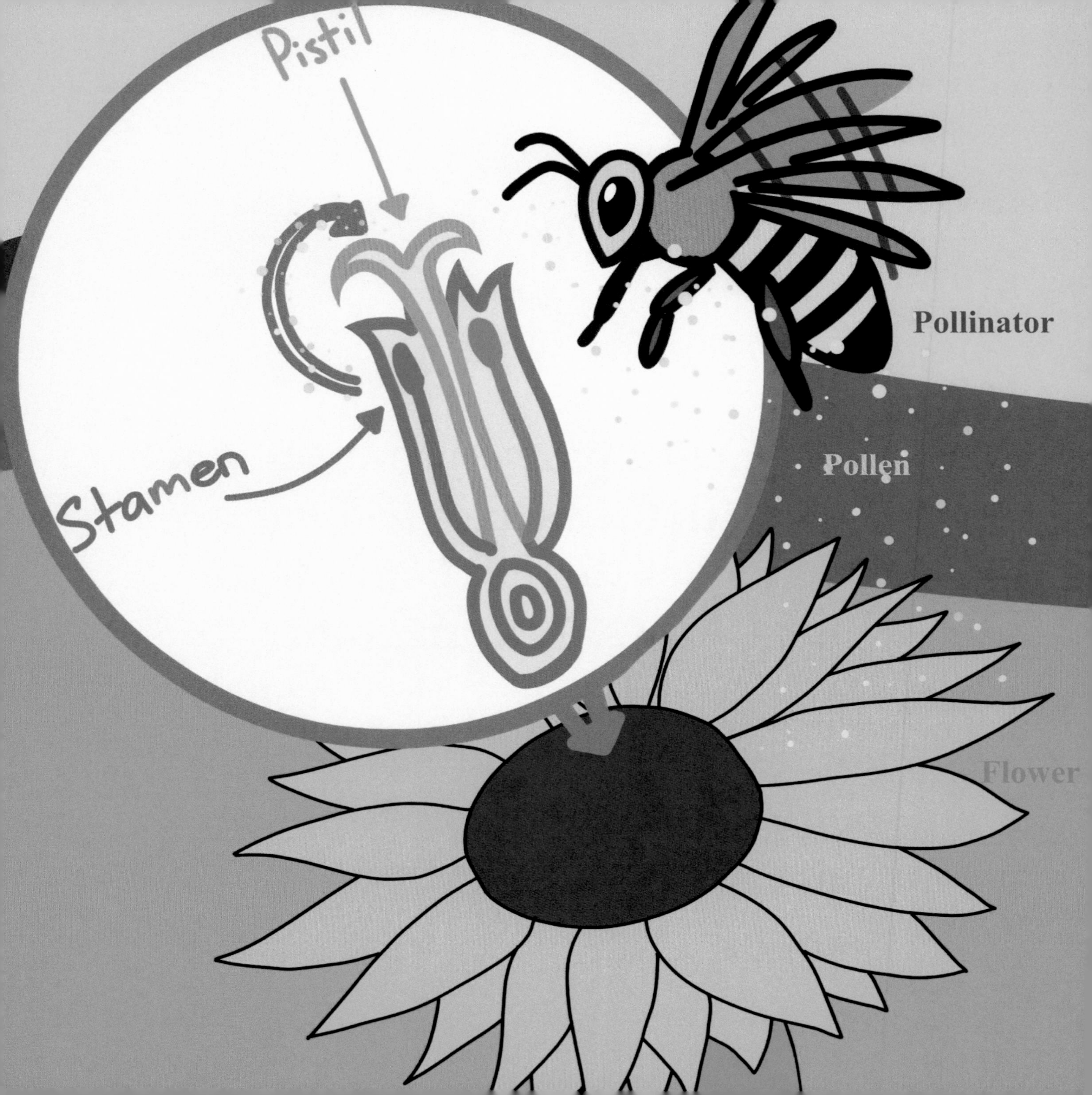
Pistil
Stamen
Pollinator
Pollen
Flower

Pollination **is the process where *pollen* moves from the male part of a flower called the *stamen* to the female part of a flower called the *pistil*. The pistil makes more seeds which are used to make more *flowers*. You can plant my seeds to make more sunflowers.**

Soon Sally lost her golden shine as she bent her head and wilted away.

Imani planted the seeds in the soil, and her garden blossomed with new beautiful sunflowers.

Imani gave sunflower bouquets to her family. She harvested sunflower seeds and shared them with her friends. She told them all about the Golden Life of Sally Sunflower.

FUN FACTS

Did you know that not all plants need soil to grow?

All plants need water and sunlight to grow.

Sunflowers are called sunflowers because they are shaped like the sun!

Sunflowers can grow between 5 to 10 feet.

Sunflowers have small rough hairs, which store the water needed and slow down evaporation.

Sunflower seeds can be eaten.

Sunflowers can be used to produce yellow dye.

The lifespan of a sunflower is between five and twelve days.

PHOTOSYNTHESIS CHART

Label the diagram to show the process of photosynthesis

GROW YOUR OWN PLANT

Name of seed planted: ____________________ Date planted: __________		
Week: __________ Drawing: Observation: ______________________ ______________________ ______________________ ______________________	Week: __________ Drawing: Observation: ______________________ ______________________ ______________________ ______________________	Week: __________ Drawing: Observation: ______________________ ______________________ ______________________ ______________________
Week: __________ Drawing: Observation: ______________________ ______________________ ______________________ ______________________	Week: __________ Drawing: Observation: ______________________ ______________________ ______________________ ______________________	Week: __________ Drawing: Observation: ______________________ ______________________ ______________________ ______________________

Made in the USA
Monee, IL
12 December 2019